A.I. For The Afrodescendants

A Practical Guide To Using, Understanding, and Mastering A.I.

Naaji Mansoor

Afrodescendant Solutions

A.I. for Afrodescendants: A Practical Guide to Understanding, Using, and Mastering A.I.

Publisher: Afrodescendant Solutions

Email: AfrodescendantSolutions@gmail.com

First Edition: 2026

ISBN: 978-1-972094-05-1

Disclaimer

Trademarks

Printed in the United States of America

This work is dedicated to Rukkiyya Mansoor,
Arifah and Halif Hasan.
in recognition of their unwavering dedication, tireless work, and meaningful contributions to the advancement and upliftment of our people. The commitment to progress and purpose stands as a constant source of inspiration.
It is also dedicated to The Honorable Silis Muhammad and Queen Misshaki Muhammad,
for their tireless and selfless efforts on behalf of our people, particularly their work within the international arena, including at the United Nations. Their dedication to justice, advocacy, and global recognition continues to pave the way for future generations.
May this work honor your sacrifices and contribute to the continued pursuit of progress, unity, and truth.

Introduction

We are living in a time of rapid transformation.

Technology is no longer something that exists in the background, it is shaping how we think, how we work, how we communicate, and how we survive. Every day, systems are becoming smarter, faster, and more integrated into our lives. From the way we search for information to the way businesses are built, technology is no longer optional, it is essential.

At the center of this transformation is Artificial Intelligence.

For many, A.I. feels complicated.

For others, it feels intimidating.

And for some, it feels like something meant for "other people", engineers, programmers, or those with advanced degrees.

But that perception is no longer true.

A.I. is not reserved for a select few.

It is not limited to experts.

And it is not something you need years of training to begin using.

It is a tool.

And like any tool, its power depends on who uses it, and how.

This book was created with a clear purpose: to remove the confusion, eliminate the fear, and provide a direct path to understanding and using Artificial Intelligence in real, practical ways. Whether you are young or old, experienced or just beginning, this guide is designed to meet you where you are and move you forward.

For too long, access to knowledge has been uneven. Systems of education, economics, and opportunity have not always been structured in ways that benefit everyone equally. Information has often been filtered, delayed, or made unnecessarily complex, creating barriers where there should have been bridges.

Artificial Intelligence has the potential to change that.

With the right approach, A.I. can:

Explain complex ideas in simple terms

Provide guidance without judgment

Help build businesses from the ground up

Offer access to knowledge that was once restricted

But it can also do the opposite.

If misunderstood or misused, A.I. can:

Spread misinformation

Reinforce existing biases

Create dependency without understanding

Replace effort with shortcuts that weaken growth

That is why understanding A.I. is no longer optional, it is necessary.

This book does not assume you have a technical background.

It does not require prior knowledge.

And it does not speak over your head.

Instead, it is structured to give you:

A clear understanding of what A.I. is

A realistic view of its benefits and risks

Step-by-step guidance on how to begin using it

Practical prompts and examples you can apply immediately

Insight into how A.I. is reshaping access, opportunity, and knowledge

Most importantly, it is written with intention.

The intention to ensure that this moment, this shift in technology, does not pass by unnoticed or unused. The intention to equip you with the tools needed not just to keep up, but to move ahead.

Because the reality is simple:

Those who understand A.I. will have an advantage.

Those who use it effectively will create opportunities.

And those who ignore it risk being left behind in a world that will not slow down.

This is not about becoming a programmer.

This is not about mastering complex systems overnight.

This is about learning how to use a powerful tool to:

Improve your life

Expand your knowledge

Strengthen your position

And contribute to something greater than yourself

You do not need permission to begin.

You do not need perfect understanding.

You only need a willingness to learn.

This book is your starting point.

And by the time you reach the final page, you will not only understand what Artificial Intelligence is—you will know how to use it, how to benefit from it, and how to move forward with clarity and purpose.

The opportunity is here.

The tools are available.

Now it is time to begin.

Table of Contents

Table of Contents

Contents

Chapter One

Chapter 1

What Is A.I.? A Simple and Clear Understanding

Artificial Intelligence, commonly known as A.I., is one of the most talked-about technologies in the world today. Yet for many, it remains misunderstood, overcomplicated, or even intimidating.

In simple terms, Artificial Intelligence is the ability of machines and computer systems to perform tasks that normally require human thinking.

That's it.

It is not magic.

It is not a replacement for human beings, in necessary terms.

It is a tool, designed to process information, recognize patterns, and generate responses based on what it has learned.

Understanding A.I. in Everyday Life

Before going any further, it is important to recognize that A.I. is already part of your daily life, even if you have never used it intentionally.

When your phone suggests the next word while typing, that is A.I.

When a video platform recommends content based on what you watch, that is A.I.

When maps suggest the fastest route to your destination, that is A.I.

Even tools like ChatGPT, Google Gemini, and Microsoft Copilot are simply more advanced forms of the same concept, systems designed to assist, respond, and generate information based on input.

A.I. is not something new being introduced into your life.

It is something that has been quietly shaping it for years.

A.I. Is Not Human Intelligence

One of the biggest misunderstandings about A.I. is the belief that it "thinks" like a human being.

It does not.

A.I. does not have:

Emotions

Consciousness

Personal experience

True understanding

Instead, it works by:

Analyzing large amounts of data

Recognizing patterns

Predicting the most likely response

When you ask A.I. a question, it is not "thinking" about the answer, it is calculating the best possible response based on patterns it has learned. (and or restraints, guidelines, or control of information, suggestions, blockages etc. it has been programmed to do. It is controlled by man).

This distinction is important.

Because it reminds us that A.I. is powerful, but it is not all-knowing, and it is not always correct. You have to monitor it closely.

The Power of A.I. as a Tool

If A.I. is not human, then what makes it so powerful?

The answer is speed and access.

A.I. can:

Process information faster than any human

Pull from vast amounts of (permitted) knowledge instantly

Simplify complex topics into understandable explanations

Assist with tasks that would normally take hours

What once required:

A teacher

A consultant

A researcher

A team

Can now begin with a single question.

This changes everything.

A.I. as an Assistant, Not a Replacement

It is important to understand the proper role of A.I.

A.I. is not necessarily here to replace your thinking, it is here to assist it.

Used correctly, A.I. becomes:

A guide

A helper

A starting point

Used incorrectly, it becomes:

A shortcut without understanding

A source of misinformation

A replacement for effort

The difference is not in the technology.

The difference is in the user!!!

Why A.I. Matters Right Now

There have been many technological shifts throughout history:

The printing press

The industrial revolution

The internet

Each one changed how people lived, worked, and communicated.

Artificial Intelligence is another one of those moments.

But there is something different about this one.

This time, the tool is:

Faster

More accessible

Easier to use

Available to almost anyone

Which means the gap between those who use it and those who do not will grow quickly.

Those who learn how to use A.I. will:

Move faster

Learn more efficiently

Create more opportunities

Those who ignore it may find themselves trying to catch up in a system that is already moving ahead.

A.I. Is a Starting Point, Not the Finish Line

One of the most important things to understand is this:

A.I. does not replace effort, it amplifies it.

If you are:

Focused → A.I. helps you move faster

Curious → A.I. helps you learn more

Disciplined → A.I. helps you execute better

But if you are:

Unfocused → A.I. will not fix that

Undisciplined → A.I. will not correct that

Passive → A.I. will not move for you

It is a multiplier, not a miracle.

Breaking the Fear Around A.I.

Many people hesitate to use A.I. because they feel:

"I'm not tech-savvy"

"I don't understand it"

"It's too advanced for me"

But the truth is:

A.I. was designed to be used by everyday people.

You do not need:

A degree

Technical training

Programming skills

You only need:

A question

A willingness to learn

A decision to start

The First Step Is the Most Important

Everything in this book begins with a simple action:

Asking your first question.

Not perfectly.

Not professionally.

Just honestly.

Because once you begin, you will realize something important:

A.I. is not something to fear.

It is something to learn.

And once learned, it becomes something you can use to improve your life in ways that were not easily possible before.

Chapter Reflection

Artificial Intelligence is not a mystery.

It is not beyond your reach.

And it is not reserved for a select few.

It is a tool, one that is now available to you.

The only question that remains is:

Will you use it?

Chapter Two

Chapter 2

Chapter 2

The History of A.I. And Who Controls It

Artificial Intelligence did not appear overnight.

It is the result of decades of research, experimentation, and development, built step by step by governments, universities, and powerful technology companies. To understand where A.I. is today, it is important to understand where it came from, and who has shaped it along the way.

The Early Idea of Artificial Intelligence

The concept of machines thinking like humans is not new. As early as the 1940s and 1950s, scientists began asking a simple but powerful question:

Can a machine be taught to think?

One of the earliest pioneers in this field was Alan Turing, who proposed that if a machine could respond to questions in a way that was indistinguishable from a human, it could be considered intelligent.

This idea laid the foundation for what would later become Artificial Intelligence.

At the time, however, computers were large, slow, and limited. The vision existed, but the technology was not yet ready.

The Rise of Computer Power

As computers became faster and more capable in the 1970s, 80s, and 90s, researchers began experimenting with systems that could:

Solve problems

Recognize patterns

Make basic decisions

These early systems were not advanced by today's standards, but they introduced a critical concept:

Machines could be trained using data.

This marked the beginning of what we now call machine learning, a core part of modern A.I.

The Internet Changed Everything

The true turning point came with the rise of the internet.

For the first time in history:

Massive amounts of information became available. Data could be collected on a global scale. Systems could learn from real human behavior.

Search engines, online platforms, and digital tools began gathering data, what people searched, what they clicked, what they watched.

This data became the fuel for A.I. The more data available, the smarter the systems could become.

Modern A.I. and Big Tech

Today's A.I. systems are far more advanced than anything imagined in the early days. They can:

Write text.

Generate images.

Translate languages.

Analyze complex problems.

Much of this progress has been driven by major technology companies such as:

Google

Microsoft

OpenAI

Meta

These organizations invest billions of dollars into A.I. research and development. They build the systems, control the platforms, and shape how A.I. is introduced to the public.

This brings us to an important question:

Who Controls A.I.?

A.I. is not controlled by a single person or group, but it is influenced heavily by those who:

Build the systems

Own the platforms

Control the data

This includes:

Large corporations

Government agencies

Research institutions

These entities decide:

What data is used

How systems are trained

What rules and limits are applied

This influence matters.

Because A.I. systems reflect the data they are trained on, and the decisions made by those who create them.

Understanding Bias in A.I.

No system is perfect.

Because A.I. learns from existing data, it can also reflect:

Historical inequalities

Cultural biases

Incomplete or inaccurate information

If the data used to train A.I. contains bias, the system may repeat or amplify it.

This is why it is important to:

Question outputs

Verify important information

Use A.I. as a tool, not as absolute truth

Understanding this does not weaken A.I.'s value, it strengthens your ability to use it wisely.

Access vs Control

Here is the key shift happening today:

While A.I. systems are developed and controlled by large organizations, access to A.I. is becoming more widely available than ever before.

Tools like ChatGPT and others allow everyday people to:

Ask questions

Generate content

Learn new skills

Solve problems

This creates a new dynamic:

Control may remain concentrated, but access is expanding.

And access changes everything.

Why This Matters for You

Understanding the history and control of A.I. is not just about technology, it is about awareness.

It allows you to:

Use A.I. without blind trust

Recognize its strengths and limitations

Avoid misinformation

Apply it strategically

Most importantly, it helps you understand that:

A.I. is not neutral.

It is shaped by human decisions.

And because of that, it must be used with intention.

The Shift We Are Living In

We are currently in a transition period.

A time where:

A.I. is rapidly improving

Access is increasing

Understanding is still catching up

Those who take the time to learn now will be ahead of the curve.

Those who ignore it may find themselves adapting later under pressure.

Chapter Reflection

Artificial Intelligence did not appear by accident.

It was built over time, shaped by powerful institutions, and fueled by data collected from the world itself.

It is influenced.

It is evolving.

And it is now accessible.

The goal is not to fear it.

The goal is to understand it, so that you can use it with awareness, purpose, and control.

Because the more you understand the system,

the less likely you are to be controlled by it.

Chapter Three

Chapter 3

The Good, The Bad, and The Ugly of A.I.

Artificial Intelligence is powerful.

But like all powerful tools, it is not automatically good or bad, it depends on how it is used, who is using it, and what purpose it serves.

A.I. can open doors that were once closed.

It can also create new challenges that were never faced before.

To truly understand A.I., you must see it from all sides:

The Good, what it makes possible

The Bad, where it creates risk

The Ugly, where it can cause real harm

Only then can you use it wisely.

The Good, Opportunity, Access, and Advancement

At its best, A.I. is a tool of empowerment.

It has the ability to break down barriers that have existed for generations, especially in areas like education, business, and access to information.

1. Access to Knowledge

What once required:

Expensive education

Specialized training

Access to institutions

Can now begin with a simple question using tools like ChatGPT.

A student struggling with a subject can now receive explanations in plain language.

An adult returning to learning can move at their own pace.

An idea can be explored instantly instead of waiting for access.

Knowledge is no longer locked, it is available.

2. Business and Economic Opportunity

A.I. allows individuals to:

Create business plans

Generate marketing content

Build brands

Automate tasks

What once required a team can now begin with one person and a clear idea.

This lowers the barrier to entry.

3. Speed and Efficiency

Tasks that once took hours, or days, can now be done in minutes.

A.I. can:

Draft documents

Organize information

Provide solutions quickly

This allows individuals to move faster and focus on higher-level thinking.

4. Personalized Learning

A.I. adapts to the user.

It can:

Explain things in different ways

Adjust to your level of understanding

Provide step-by-step guidance

This creates a learning experience that is more flexible than traditional systems.

The Bad, Risks, Misuse, and Dependency

While A.I. creates opportunity, it also introduces risk, especially when used without awareness.

1. Over-Reliance

One of the biggest dangers is depending on A.I. too much.

If you:

Stop thinking critically

Accept answers without question

Use A.I. as a replacement for effort

You weaken your own ability to think and solve problems.

A.I. should assist your thinking, not replace it.

2. Misinformation

A.I. is not always correct.

It can:

Provide incomplete answers

Misinterpret questions

Generate incorrect information

If users do not verify what they receive, misinformation can spread quickly.

3. Job Disruption

A.I. is changing how work is done.

Some roles may:

Evolve

Become automated

Require new skills

This does not mean opportunity disappears, but it does mean adaptation is necessary.

Those who learn A.I. can benefit.

Those who ignore it may struggle to adjust.

4. Data and Privacy Concerns

A.I. systems rely on data.

This raises important questions:

What data is being collected?

How is it being used?

Who has access to it?

Understanding this helps you use A.I. more responsibly.

The Ugly, When Power Is Misused

This is the side of A.I. that must be taken seriously.

Because when powerful tools are misused, the impact can be significant.

1. Bias and Inequality

A.I. systems learn from data, and that data reflects the world as it is.

If the data contains:

Bias

Inequality

Historical imbalance

Then A.I. can repeat or amplify those patterns.

This is why awareness matters.

2. Manipulation and Influence

A.I. can be used to:

Generate misleading content

Influence opinions

Spread false narratives

In a digital world, information moves quickly, and not all of it is accurate.

Learning how to question and verify information becomes critical.

3. Surveillance and Control

A.I. can also be used to monitor behavior:

Tracking activity

Analyzing patterns

Predicting actions

While this can be used for safety and efficiency, it can also raise concerns about privacy and control.

4. Deepfakes and False Reality

A.I. can now create:

Fake images

Fake videos

Fake voices

These can be difficult to distinguish from reality.

This creates a new challenge:

Not everything you see or hear can be trusted at face value.

Balance, The Key to Using A.I. Correctly

A.I. is not something to blindly trust.

It is not something to completely reject.

It is something to understand and use with balance.

Use A.I. to:

Learn

Build

Improve

But also:

Think for yourself

Verify information

Stay aware of its limits

The Responsibility of the User

At the center of all of this is one truth:

The power of A.I. does not come from the system itself.

It comes from the person using it.

You decide:

Whether it becomes a tool of growth

Or a source of confusion

You decide:

Whether it strengthens your thinking

Or replaces it

You decide:

Whether it is used responsibly

Or carelessly

Chapter Reflection

Artificial Intelligence is not one thing, it is many things at once.

It is:

Opportunity

Risk

Power

The Good shows what is possible.

The Bad shows what to watch for.

The Ugly shows what must be handled with care.

Understanding all three is what separates a user from a master.

Because those who only see the good may be unprepared.

Those who only see the bad may miss opportunity.

But those who understand all sides

are the ones who can use A.I. with clarity, purpose, and control.

Chapter Four

Chapter 4

Chapter 4

Where to Start with A.I. Your First Step Into the System

At this point, you understand what A.I. is.

You understand its power, and its risks.

Now comes the most important part:

Starting!!!

Not perfectly. Not with full knowledge. But simply taking the first step.

Because the difference between those who benefit from A.I. and those who do not is not intelligence, it is action.

A.I. Is Easier Than You Think

Many people hesitate because they believe A.I. is complicated.

They imagine:

Coding

Technical systems

Advanced knowledge

But modern A.I. tools are designed for everyday use.

If you can:

Type a message

Ask a question

Use a phone or computer

Then you can use A.I.

Step 1: Choose Your First A.I. Tool

You do not need to use everything.

You only need to start with one.

Here are some of the most accessible tools:

ChatGPT — Best for writing, learning, and general use

Google Gemini — Good for research and explanations

Microsoft Copilot — Useful for productivity and documents

Canva — Great for design and visuals

Recommendation: Start with ChatGPT

It is simple, flexible, and powerful.

Step 2: Create an Account

Getting started is straightforward:

Go to the platform

Sign up using your email

Choose a free or paid plan

Important:

You do not need to pay to begin.

The free version is enough to learn and grow.

Step 3: Understand How A.I. Works (In Practice)

A.I. works based on something called a prompt.

A prompt is simply:

The instruction or question you give to the A.I.

Example:

"Explain what A.I. is"

"Help me start a business"

"Create a workout plan"

The quality of your result depends on:

What you ask

How clearly you ask it

Step 4: Ask Your First Question

This is where everything begins.

Do not overthink it.

Start simple:

"What is A.I. in simple terms?"

"How can I make extra money?"

"Help me organize my day"

The goal is not perfection, the goal is interaction.

Step 5: Learn Through Use

A.I. is not something you study first and use later.

You learn it by:

Asking questions

Testing ideas

Trying different prompts

Each time you use it, you get better.

Your First Real Exercise

Right now, take action.

Open ChatGPT and type:

"Create a simple plan to improve my life in the next 30 days."

Read the response.

Then ask:

"Make it simpler."

"Break it down step-by-step."

"Adjust it for my schedule."

This is how you begin to understand the system.

Common Beginner Mistakes (Avoid These Early)

1. Being Too Vague

"Help me"

"Help me create a business plan for a clothing brand"

2. Expecting Perfect Answers Immediately

A.I. improves as you refine your questions.

3. Not Asking Follow-Up Questions

A.I. is a conversation, not a one-time response.

4. Giving Up Too Quickly

The first interaction may feel basic, that is normal.

Keep going.

Step 6: Start Using A.I. in Daily Life

You do not need a major project to begin.

Use A.I. for:

Planning your day

Learning new topics

Writing messages

Solving small problems

The more you use it, the more natural it becomes.

Step 7: Build Confidence Through Repetition

Confidence does not come from reading, it comes from doing.

Every time you:

Ask a better question

Get a useful answer

Apply what you learn

You are building a skill.

And that skill compounds over time.

A.I. Is a Tool You Grow Into

You do not need to master everything at once.

Start simple.

Stay consistent.

Build gradually.

What feels new today will feel natural tomorrow.

The Real Barrier Is Not Technology

It is hesitation.

Waiting until:

You feel ready

You understand everything

You have the "perfect" starting point

That moment does not come.

Progress begins when you decide to start anyway.

Chapter Reflection

Starting with A.I. is not complicated.

It requires:

A tool

A question

A willingness to try

That is all.

Everything else builds from there.

The tools are available.

The access is open.

The barrier is low.

Now the only question is:

Will you begin?

Chapter Five

Chapter 5

Chapter 5

Step-by-Step Mastery How to Use A.I. Like a Pro

Starting is important.

But knowing how to use A.I. effectively, that is where the real power is.

Many people try A.I. once or twice, get basic results, and stop.

Not because A.I. isn't useful, but because they never learned how to use it properly.

This chapter will change that.

You are going to learn how to move from:

Basic user → Strategic user

Asking questions → Getting results

Trying A.I. → Using A.I. with purpose

The Truth About A.I. Results

A.I. gives back what you give it.

Weak input → weak output

Clear input → strong output

Detailed input → powerful output

This is the key:

The quality of your prompt determines the quality of your result.

Step 1: Think Before You Type

Before asking A.I. anything, ask yourself:

What do I want?

What outcome am I looking for?

What details matter?

The clearer you are, the better A.I. performs.

Step 2: Use the "Role + Task + Detail" Formula

This is one of the most powerful ways to use A.I.

Formula:

Act as a [role]. Help me [task]. Provide [details].

Example 1 (Basic):

"Help me start a business"

Example 2 (Advanced):

"Act as a business consultant. Help me create a step-by-step plan to start a food truck business with a $10,000 budget. Include startup costs, marketing strategy, and daily operations."

Same idea.

Completely different results.

Step 3: Be Specific

A.I. cannot read your mind.

The more specific you are, the more useful the answer becomes.

Weak Prompt:

"Give me advice"

Strong Prompt:

"Give me 5 practical ways to improve my time management while working 10-hour days"

Specificity creates clarity.

Clarity creates results.

Step 4: Use Follow-Up Questions

This is where most beginners fail.

They ask one question, and stop.

Professionals keep going.

Example Flow:

"Create a business plan"

"Make it simpler"

"Break it into steps"

"What mistakes should I avoid?"

"How can I market this for free?"

Now you are not just using A.I.

You are extracting value from it

Step 5: Ask for Different Formats

A.I. can present information in many ways.

Try asking for:

Step-by-step instructions

Bullet points

Paragraph explanations

Checklists

Schedules

Example:

"Turn this into a step-by-step checklist I can follow daily"

This makes information easier to apply.

Step 6: Simplify or Expand

If something is too complex:

"Explain this in simple terms"

If something is too basic:

"Go deeper and give more detail"

You are in control of the depth.

Step 7: Apply What You Get

This is where mastery separates from curiosity.

A.I. is not meant to be read only, it is meant to be used.

If A.I. gives you:

A plan → take action

A strategy → test it

Advice → apply it

Without action, A.I. becomes entertainment.

With action, it becomes transformation.

Step 8: Build Systems, Not Just Answers

Instead of asking random questions, begin to think in systems.

Example:

Instead of:

"Give me a workout plan"

Ask:

"Create a weekly workout system I can follow for the next 90 days with progression and tracking"

This shifts you from:

Short-term thinking → long-term strategy

Step 9: Use A.I. Consistently

Mastery comes from repetition.

Use A.I. daily for:

Learning

Planning

Problem solving

Creating

The more you use it, the sharper you become.

Step 10: Combine A.I. With Your Thinking

A.I. should not replace your ideas, it should enhance them.

You bring:

Experience

Judgment

Vision

A.I. brings:

Speed

Structure

Support

Together, they create something more powerful.

The Difference Between a Beginner and a Pro

A beginner:

Asks simple questions

Accepts basic answers

Uses A.I. occasionally

A professional:

Gives clear instructions

Refines responses

Uses A.I. daily

Applies results

The difference is not intelligence.

It is approach.

Your Practice Exercise

Open ChatGPT and type:

"Act as a life coach. Help me create a structured daily routine that improves my discipline, productivity, and health. Break it into morning, afternoon, and evening."

Then follow up with:

"Make it simpler"

"Adjust it for someone working long hours"

"Turn this into a checklist"

This is how mastery begins.

Chapter Reflection

Using A.I. like a professional is not about knowing everything.

It is about:

Asking better questions

Refining your approach

Taking action

A.I. is not the advantage.

Knowing how to use it is.

And once you understand that, you are no longer just a user,

You are in control.

Chapter Six

Chapter 6

Chapter 6

Beginner Prompts How to Talk to A.I. and Get Real Results

Now that you understand what A.I. is and how to start using it, the next step is learning how to communicate with it effectively.

This is where most people either unlock real value, or stay stuck with basic results.

The key to using A.I. is simple:

A.I. responds based on what you ask, and how you ask it.

Your words matter.

Your clarity matters.

Your intention matters.

This chapter will show you how to ask better questions and get better answers.

What Is a Prompt?

A prompt is the instruction or question you give to A.I.

It can be:

A question

A request

A task

Examples:

"Explain A.I. in simple terms"

"Create a workout plan"

"Help me start a business"

Every result you get begins with a prompt.

Why Prompts Matter

Think of A.I. like a tool that follows direction.

If your direction is:

Unclear → the result will be unclear

Basic → the result will be basic

Detailed → the result will be powerful

Start Simple Then Improve

You do not need perfect prompts to begin.

Start simple:

"Help me create a daily routine"

Then improve it:

"Create a daily routine for someone working 10-hour days who wants to improve health and productivity"

Same idea, better result.

Beginner Prompts You Can Use Right Now

These are practical prompts you can copy and use immediately.

Everyday Life

"Create a simple weekly plan to improve my daily habits"

"Help me organize my day more efficiently"

"Give me 5 ways to reduce stress and stay focused"

Business & Money

"Create a step-by-step plan to start a small business with little money"

"Give me 10 side hustle ideas I can start immediately"

"Write a marketing post for my product or service"

Education & Learning

"Explain this topic in simple terms like I'm a beginner"

"Summarize this information into key points"

"Create a study plan for me to learn this subject in 7 days"

Health & Fitness

"Create a 3-day workout plan for someone with a busy schedule"

"Give me a healthy meal plan for the week"

"Help me build better daily habits for energy and focus"

Personal Growth

"Help me build a daily discipline routine"

"Give me steps to improve my confidence"

"Help me set goals and achieve them step-by-step"

How to Improve Any Prompt

If your result is not strong, improve your prompt using these methods:

1. Add Detail

"Help me make money"

"Give me 5 ways to make extra money from home with little startup cost"

2. Add Context

"Create a plan"

"Create a daily plan for someone working full-time who wants to improve productivity"

3. Ask for Structure

"Explain this"

"Explain this step-by-step in simple terms"

4. Ask for Simplicity

"Make this easier to understand"

5. Ask for Expansion

"Go deeper and give more detail"

The Power of Follow-Up Prompts

One of the most important skills is continuing the conversation.

Example:

"Create a business plan"

"Make it simpler"

"Break it into steps"

"What mistakes should I avoid?"

"How can I market this for free?"

This is how you turn one answer into real value.

The Beginner Mistake to Avoid

Many people:

Ask one question

Read one answer

Stop

That is not how A.I. is meant to be used.

A.I. is a conversation.

The more you interact, the better your results become.

Your First Prompt Practice

Open ChatGPT and type:

"Help me create a plan to improve my life over the next 30 days. Make it simple and easy to follow."

Then follow up with:

"Break it into daily steps"

"Adjust it for my schedule"

"Turn this into a checklist"

Confidence Comes From Use

At first, your prompts may feel basic.

That is normal.

With practice:

You will ask better questions

You will get better answers

You will move faster

A.I. Understands More Than You Think

You do not need perfect wording.

You can:

Speak naturally

Ask directly

Keep it simple

The key is to start and improve over time.

Chapter Reflection

A.I. is not complicated, it is responsive.

It listens to what you ask.

It responds to how you ask.

And it improves as you improve.

The skill is not in the system.

The skill is in the question.

And once you learn how to ask better questions, you unlock better answers—and better outcomes.

Chapter Seven

Chapter 7

Chapter 7

Prompt Mastery How to Get Elite-Level Results from A.I.

By now, you understand what a prompt is and how to use basic instructions.

This chapter is where you move beyond beginner level.

This is where you learn how to use A.I. with precision, control, and strategy.

Because the difference between average results and powerful results is not the tool,

It is the level of mastery in how you communicate with it.

From Basic User to Strategic User

A beginner asks:

"Help me with a business idea"

A skilled user asks:

"Act as a business strategist. Give me 3 scalable business ideas in the food industry with low startup cost, high demand, and clear profit potential. Include pros, cons, and first steps."

Same tool.

Different result.

The Master Prompt Formula

This is your foundation:

Role + Task + Context + Format + Depth

1. Role Who should A.I. act as?

This gives the response direction.

Examples:

"Act as a business consultant"

"Act as a fitness coach"

"Act as a financial advisor"

2. Task What do you want done?

Be clear and direct.

Examples:

"Help me create a business plan"

"Build a workout routine"

"Explain this topic"

3. Context Your situation

This personalizes the result.

Examples:

Budget

Time constraints

Experience level

4. Format How should it be delivered?

Examples:

Step-by-step

Checklist

Bullet points

Daily schedule

5. Depth How detailed should it be?

Examples:

"Keep it simple"

"Go deep and be detailed"

Full Example of a Master Prompt

"Act as a business consultant. Help me create a step-by-step plan to start a clothing brand with a $2,000 budget. Include startup costs, marketing strategy, and daily actions. Keep it simple but detailed enough to follow."

This is how professionals use A.I.

Stacking Prompts (Advanced Technique)

Do not stop at one prompt.

Build on it.

Example:

"Create a business plan"

"Now create a marketing strategy for this plan"

"Now give me a 30-day action plan"

"What mistakes should I avoid?"

This creates depth and structure.

Using Constraints for Better Results

Constraints make A.I. more precise.

Examples:

"Under $1,000 budget"

"For beginners only"

"For someone working full-time"

Constraints = more realistic results

Ask A.I. to Think Step-by-Step

This improves clarity.

Example:

"Break this down step-by-step so I can follow it easily"

This forces A.I. to organize its response clearly.

Refining Responses (Professional Skill)

Your first answer is not your final answer.

Refine it.

Use these prompts:

"Make this simpler"

"Expand on this section"

"Give more examples"

"Rewrite this in a clearer way"

This is how you improve output quality.

Turning A.I. Into a System Builder

A beginner asks for answers.

A master builds systems.

Example:

"Give me a workout plan"

"Create a 90-day fitness system with weekly progression, tracking, and adjustments"

Systems create long-term results.

Avoid These Common Mistakes

1. Being Too General

You get general answers.

2. Not Using Follow-Ups

You leave value on the table.

3. Not Applying Results

Knowledge without action = no change.

4. Expecting Perfection

Refinement is part of the process.

Using A.I. for Strategy

A.I. is not just for tasks—it is for thinking.

You can ask:

"What is the best strategy for this situation?"

"Compare 3 options and give pros and cons"

"What would an expert do in this situation?"

This elevates your thinking.

The Real Advantage

Most people will:

Use A.I. casually

Ask basic questions

Get average results

Very few will:

Master prompts

Build systems

Use A.I. strategically

That is where the advantage is.

Your Mastery Exercise

Open ChatGPT and type:

"Act as a life strategist. Help me build a 30-day plan to improve my discipline, finances, and health. Break it into daily steps with clear actions and realistic goals."

Then refine:

"Make it simpler"

"Adjust it for a busy schedule"

"Turn it into a checklist"

This is how mastery begins.

Chapter Reflection

A.I. is not just a tool, it is a multiplier.

It multiplies:

Your clarity

Your thinking

Your action

The more precise you become,

the more powerful your results become.

Because in the end:

The system is only as strong as the person using it.

And once you master the way you communicate with A.I.,

you are no longer just asking questions,

You are directing outcomes.

Chapter Eight

Chapter 8

Chapter 8

Making Money with A.I. Turning Knowledge Into Income

Artificial Intelligence is not just a tool for learning, it is a tool for earning.

For the first time, individuals can take ideas, skills, and information and turn them into income faster and with fewer barriers than ever before.

What once required:

Large startup capital

Teams of people

Specialized knowledge

Can now begin with:

A laptop or phone

An internet connection

A clear idea

And the willingness to act.

A New Economic Opportunity

A.I. changes how money can be made.

It allows you to:

Start faster

Work smarter

Create more with less

This does not mean money comes easily.

It means the entry point is lower, and the opportunity is greater.

The Key Principle

Before going further, understand this:

A.I. does not make money for you. It helps you create value, and value creates income.

If you focus on:

Solving problems

Providing services

Creating useful content

Then A.I. becomes a powerful advantage.

1. Freelancing with A.I.

One of the fastest ways to start making money is by offering services.

A.I. can help you provide:

Writing services

Social media content

Resume creation

Business plans

Using tools like ChatGPT, you can:

Generate ideas

Draft content

Improve quality

Example Prompt:

"Act as a professional writer. Create a high-quality social media post for a small business promoting a new product."

You can then:

Edit it

Customize it

Sell it as a service

2. Content Creation

Content is one of the most valuable assets in today's digital world.

A.I. can help you create:

Social media posts

Blog articles

Video scripts

Captions and descriptions

Platforms like TikTok and Facebook allow you to:

Reach audiences

Build influence

Monetize attention

Example Prompt:

"Give me 10 viral content ideas about self-improvement and discipline"

Content leads to:

Followers

Engagement

Opportunities

3. Digital Products

A.I. makes it easier to create products you can sell repeatedly.

Examples:

E-books

Guides

Templates

Courses

You can use A.I. to:

Outline content

Write drafts

Organize information

Example Prompt:

“Help me create an outline for a beginner guide on budgeting and saving money”

Once created, digital products can generate income over time.

4. Business Ideas and Planning

A.I. can help you:

Generate business ideas

Create strategies

Identify opportunities

Example Prompt:

“Give me 5 business ideas I can start with under $1,000 and explain how to begin”

This allows you to move from:

Thinking → Planning → Execution

5. Automation and Efficiency

A.I. helps you do more in less time.

You can:

Automate writing

Streamline tasks

Reduce manual work

This allows you to:

Focus on growth

Serve more people

Increase productivity

6. Learning Skills That Pay

A.I. is not just for earning directly, it helps you learn skills that lead to income.

You can learn:

Marketing

Writing

Business strategy

Communication

Example Prompt:

"Teach me the basics of digital marketing in simple terms with steps I can follow"

The faster you learn, the faster you can earn.

7. Turning Knowledge Into Value

The real goal is not just using A.I., it is using it to create value.

Ask yourself:

What problems can I solve?

What skills can I offer?

What information can I organize and share?

A.I. helps you:

Structure your ideas

Improve your output

Deliver value faster

Avoid the Shortcut Mentality

One of the biggest mistakes is trying to use A.I. as a shortcut to easy money.

That mindset leads to:

Low-quality work

No long-term success

Frustration

Instead, focus on:

Quality

Consistency

Improvement

The Opportunity Is Real

Right now, many people are:

Learning A.I.

Using it daily

Building income streams

At the same time, many are:

Ignoring it

Hesitating

Waiting

The difference between these two groups will grow over time.

Your First Income Exercise

Open ChatGPT and type:

"Act as a business coach. Give me 3 ways I can start making money using A.I. with little to no experience. Include step-by-step instructions."

Then follow up with:

"Which one is easiest to start?"

"Break it into daily steps"

"What mistakes should I avoid?"

This is how you begin.

Chapter Reflection

Artificial Intelligence does not remove the need for effort, it increases the return on effort.

It allows you to:

Start sooner

Learn faster

Create more

But it still requires:

Action

Discipline

Consistency

The opportunity is not hidden.

It is available.

And those who take advantage of it will not just keep up,

They will move ahead.

Chapter Nine

Chapter 9

Chapter 9

A.I. for Students and Youth Learning Without Limits

For generations, learning has followed a fixed path.

A classroom.

A teacher.

A schedule.

A pace that not everyone could keep up with.

If you understood quickly, you moved forward.

If you struggled, you were often left behind.

That system created limits.

But today, those limits are being challenged.

Artificial Intelligence introduces a new way of learning—one that is flexible, personal, and always available.

A way to learn without limits.

A New Era of Learning

A.I. transforms learning from passive to active.

Instead of:

Waiting to be taught

You can now:

Ask

Explore

Adjust

Repeat

At any time.

Using tools like ChatGPT, students can turn confusion into understanding instantly.

Learning at Your Own Pace

Not everyone learns the same way.

Some need:

More time

More examples

Simpler explanations

A.I. adapts to you.

You can say:

"Explain this step-by-step"

"Make it simpler"

"Give another example"

And it will adjust.

This creates something powerful:

A personalized learning experience

Breaking Down Difficult Subjects

Subjects that once felt overwhelming, such as math, science, or writing, can now be approached differently.

Instead of struggling alone, a student can ask:

"Explain this like I'm a beginner"

"Walk me through this step-by-step"

"Test me so I can learn it better"

A.I. does not rush.

It does not judge.

It does not get tired of explaining.

A.I. as a 24/7 Tutor

Imagine having:

A tutor available anytime

Unlimited patience

The ability to adjust explanations instantly

That is what A.I. provides.

Whether it is early morning or late at night, help is available.

Using A.I. the Right Way for School

A.I. can support learning, but it must be used correctly.

Used properly, it helps you:

Understand concepts

Practice problems

Improve your thinking

Used incorrectly, it can:

Do the work for you

Prevent real learning

Create dependency

The goal is not to copy answers.

The goal is to understand.

Study Smarter, Not Harder

A.I. helps organize your learning.

You can ask:

"Create a 7-day study plan for my test"

"Summarize this topic into key points"

"Quiz me on this subject"

This turns studying into a structured system instead of guessing what to focus on.

Building Skills for the Future

Students today are not just preparing for tests, they are preparing for life.

A.I. helps develop:

Communication skills

Problem-solving ability

Critical thinking

Creativity

It can also introduce:

Business ideas

Career paths

Opportunities that were not easily visible before

Creativity and Exploration

A.I. is not just for schoolwork.

It can be used to:

Write stories

Create ideas

Build projects

Explore interests

This allows young people to:

Discover talents

Build confidence

Express themselves

The Advantage of Starting Early

Those who begin using A.I. now will:

Learn faster

Adapt quicker

Gain skills earlier

They will not be trying to catch up later.

They will already be prepared.

Guidance for Parents and Guardians

A.I. is a tool, but guidance is still important.

Encourage:

Curiosity

Learning

Exploration

But also teach:

Responsibility

Discipline

Critical thinking

Balance is key.

Responsibility Comes With Access

With access to powerful tools comes responsibility.

Students must learn to:

Verify information

Think independently

Use A.I. honestly

Because shortcuts may feel helpful in the moment,

But they weaken long-term growth.

Chapter Reflection

Artificial Intelligence has changed what is possible for students.

It provides:

Access

Support

Flexibility

But it does not replace effort.

The students who benefit most will not be those who use A.I. to avoid learning

They will be those who use it to understand more deeply and grow more quickly.

Because in this new era:

The advantage belongs to those who know how to learn.

Chapter Ten

Chapter 10

Chapter 10

A.I. for Elders Making Technology Simple, Useful, and Accessible

Technology can feel overwhelming.

For many elders, it can seem like the world has changed too quickly, new devices, new systems, new ways of doing things. What was once simple has become complex, and what was once familiar has become unfamiliar.

But Artificial Intelligence is different.

At its core, A.I. is not about complexity, it is about simplicity.

It allows you to:

Ask questions in plain language

Receive clear answers

Get help without needing technical knowledge

You do not need to understand technology to benefit from A.I.

You only need to know how to ask.

A Tool That Works Like Conversation

One of the greatest strengths of A.I. is that it works like a conversation.

Using tools like ChatGPT, you can simply type or speak as if you are talking to a person.

Examples:

"Explain this in simple terms"

"Help me understand this bill or document"

"What should I do in this situation?"

There is no special language required.

Just speak naturally.

No Experience Needed

You do not need:

Computer training

Technical skills

Prior experience

If you can:

Use a phone

Send a message

Ask a question

You can use A.I.

Practical Ways Elders Can Use A.I.

A.I. is not just for young people or professionals, it can be useful in everyday life.

1. Understanding Information

Many documents today can be confusing:

Medical information

Legal documents

Financial terms

You can ask:

"Explain this in simple terms so I can understand it"

A.I. can break complex information into clear, simple language.

2. Health and Wellness Support

While A.I. is not a doctor, it can help you:

Understand general health topics

Learn about nutrition

Get suggestions for staying active

3. Daily Organization

You can ask A.I. to help you:

Plan your day

Create reminders

Organize tasks

Example:

"Help me create a simple daily routine to stay active and productive"

4. Communication Help

A.I. can assist with:

Writing messages

Responding to emails

Expressing thoughts clearly

Example:

"Help me write a message to my family about..."

5. Learning New Things

It is never too late to learn.

You can explore:

History

Skills

Interests

At your own pace, without pressure.

Removing the Fear of Technology

One of the biggest barriers is not ability, it is fear.

The fear of:

Doing something wrong

Not understanding

Feeling left behind

But A.I. is designed to help, not judge.

You can:

Ask the same question multiple times

Request simpler explanations

Take your time learning

There is no pressure.

Step-by-Step: How to Start

Here is a simple way to begin:

Open a tool like ChatGPT

Type a simple question

Read the response

Ask a follow-up question

That is all.

You do not need to know everything.

You just need to begin.

The Role of Family and Community

Younger family members can play an important role by:

Helping elders get started

Showing how to ask questions

Encouraging consistent use

This creates:

Connection

Learning across generations

Shared growth

Technology as a Bridge, Not a Barrier

Instead of seeing technology as something that separates generations, A.I. can bring them together.

Youth can guide

Elders can learn

Knowledge can be shared

This builds stronger families and communities.

Patience and Progress

Learning something new takes time.

There is no need to rush.

Each step forward, no matter how small, is progress.

The Opportunity for Elders

A.I. provides something valuable:

Access to information

Support without barriers

The ability to stay connected and informed

It ensures that no one is left behind simply because the world has changed.

Chapter Reflection

Artificial Intelligence is not just for the young, it is for everyone.

It is:

Simple

Accessible

Useful

It removes barriers and replaces them with opportunity.

All it requires is:

Curiosity

A willingness to try

A simple question

Because it is never too late to learn.

And it is never too late to benefit from something new.

Chapter Eleven

Chapter 11

Chapter 11

A.I. in the Community Building Together, Growing Together

Artificial Intelligence is powerful on an individual level.

It can help one person learn faster, work smarter, and create new opportunities.

But its true power is revealed when it is used collectively.

When individuals within a community begin to use A.I. together, something greater happens:

Knowledge spreads faster.

Skills increase.

Opportunities multiply.

This is where A.I. moves beyond a tool, and becomes a force for community growth.

From Individual Use to Collective Power

One person using A.I. can improve their life.

But a group of people using A.I. can:

Build systems

Share knowledge

Strengthen each other

Instead of learning in isolation, communities can begin to:

Teach one another

Solve problems together

Create opportunities that benefit many

Building Smarter Communities

A.I. can help communities organize and grow more effectively.

It can be used to:

Plan events

Develop programs

Create educational materials

Improve communication

Example:

"Help me create a community program that teaches financial literacy to young people"

A.I. can provide:

Structure

Ideas

Step-by-step plans

Supporting Local Businesses

Communities thrive when businesses grow.

A.I. can support local businesses by helping with:

Marketing

Branding

Customer communication

Business planning

A small business owner can ask:

"Create a marketing plan for my local business with a limited budget"

This allows businesses to:

Compete more effectively

Reach more customers

Grow sustainably

Teaching A.I. Within the Community

One of the most powerful things a community can do is teach A.I. to others.

Imagine:

Workshops

Group learning sessions

Mentorship programs

Where individuals learn how to:

Ask better questions

Use A.I. for real-life situations

Apply knowledge immediately

This creates a ripple effect.

One person learns.

Then teaches another.

Then another.

Bridging the Generational Gap

A.I. creates an opportunity to bring generations together.

Youth often adapt quickly

Elders bring wisdom and experience

Together:

Youth can guide in technology

Elders can guide in life experience

This exchange strengthens the community.

Solving Community Problems

Every community faces challenges.

A.I. can help analyze and approach solutions.

Example:

"What are practical ways to improve economic opportunities in my community?"

A.I. can provide:

Ideas

Strategies

Different perspectives

While it does not replace real-world action, it helps guide thinking.

Creating Shared Resources

Communities can use A.I. to build:

Guides

Educational materials

Training programs

Business directories

These resources can be shared and reused, creating long-term value.

Reducing Barriers Together

Many barriers exist within communities:

Lack of access

Limited resources

Information gaps

A.I. helps reduce these barriers by:

Providing knowledge

Offering guidance

Supporting skill development

When used collectively, this impact is multiplied.

Responsibility Within the Community

With access to A.I. comes responsibility.

Communities must:

Promote ethical use

Encourage learning, not shortcuts

Share accurate information

This ensures that A.I. is used to build, not weaken, the community.

The Power of Unity and Technology

Technology alone does not create progress.

People do.

But when people come together with the right tools, progress accelerates.

A.I. provides:

Speed

Access

Support

Community provides:

Strength

Direction

Purpose

Together, they create something powerful.

A Vision for the Future

Imagine a community where:

Knowledge is shared openly

Skills are taught freely

Businesses support each other

Individuals grow together

A.I. makes this more possible than ever before.

Your Role in the Community

You do not need to wait for others to begin.

You can:

Learn

Share

Teach

Build

Even small actions create impact.

Chapter Reflection

Artificial Intelligence is not just a personal tool, it is a community tool.

When used together, it can:

Strengthen connections

Expand opportunities

Accelerate growth

The question is not whether A.I. can help communities.

The question is:

Will communities come together and use it?

Because when knowledge is shared,

growth is multiplied.

And when growth is multiplied,

the future changes.

Chapter Twelve

Chapter 12

Chapter 12

Breaking Barriers with A.I. Unlocking What Was Once Restricted

For much of history, access has been controlled.

Access to:

Knowledge

Education

Opportunity

Systems of power

Not everyone had the same entry point.

Not everyone had the same support.

And not everyone had the same chance.

Barriers existed, and in many cases, they were not accidental.

They were built through:

Cost

Complexity

Limited access

Institutional control

For many, advancement required:

The right connections

The right environment

The right resources

Without those, progress was slowed—or stopped entirely.

But something is changing.

A Shift in Access

Artificial Intelligence is not just another tool.

It is a shift in access.

It allows individuals to:

Reach information instantly

Learn without permission

Build without waiting

Understand without gatekeepers

What once required approval can now begin with a question.

Removing Educational Barriers

Education has often been tied to:

Institutions

Cost

Location

Availability

A.I. changes this.

With tools like ChatGPT, individuals can:

Learn new subjects

Understand complex topics

Practice skills

Without:

Enrollment

Tuition

Formal barriers

This does not replace education, it expands access to it.

Breaking Down Complex Knowledge

Many systems have relied on complexity to limit access.

Legal language.

Financial systems.

Technical fields.

A.I. can translate complexity into clarity.

You can ask:

"Explain this in simple terms"

And receive understanding that was once difficult to obtain.

Reducing Financial Barriers

Starting a business, learning a skill, or building something once required:

Money

Tools

Teams

Now, A.I. allows individuals to:

Plan

Create

Organize

With minimal cost.

This lowers the barrier to entry.

Expanding Opportunity

Opportunity is no longer limited to:

Location

Background

Traditional pathways

With A.I., individuals can:

Learn skills online

Build digital products

Reach global audiences

This expands what is possible.

Access Without Permission

One of the most powerful changes is this:

You no longer need permission to begin.

You do not need:

Approval

Formal entry

Gatekeepers

You can:

Ask

Learn

Apply

Immediately.

The Equalizing Effect

A.I. does not make everything equal.

But it moves things closer.

It creates:

More access

More opportunity

More possibility

The difference now becomes:

Who uses it

How it is used

How consistently it is applied

From Limitation to Possibility

Before:

Information was harder to reach

Learning required access

Opportunities were restricted

Now:

Information is immediate

Learning is flexible

Opportunities can be created

This is a shift from limitation to possibility.

Responsibility Within Access

Access alone is not enough.

It must be used.

A.I. provides:

Tools

Knowledge

Direction

But it still requires:

Action

Discipline

Consistency

Without those, access does not turn into progress.

The New Reality

We are entering a time where:

Knowledge is more available than ever

Tools are more accessible than ever

Opportunities are more flexible than ever

This changes the landscape.

The Role of the Individual

At the center of this shift is the individual.

You now have:

More access than previous generations

More tools than ever before

More ability to learn and build

What you do with it matters.

Chapter Reflection

Artificial Intelligence is not just advancing technology.

It is changing access.

It is removing barriers that once slowed progress.

It is unlocking doors that were once closed.

But unlocking a door is only the first step.

You still have to walk through it.

Because in this new environment:

Opportunity is no longer reserved, it is available.

And the difference will not be who has access.

The difference will be who uses it.

Chapter Thirteen

Chapter 13

Chapter 13

The New Level Playing Field Opportunity in a Changing World

For generations, the playing field was not level.

Access to:

Knowledge

Opportunity

Resources

Networks

Was uneven.

Some began ahead.

Some had support.

Some had systems designed to help them move forward.

Others did not.

The difference was not always effort.

The difference was often access.

A Shift Is Taking Place

Artificial Intelligence is changing that structure.

Not completely, but significantly.

It is creating a new environment where:

Information is easier to access

Tools are widely available

Learning can begin instantly

This does not erase all inequality.

But it reduces one of the most important barriers:

The barrier to knowledge.

Before A.I.

Before the rise of A.I., gaining knowledge often required:

Formal education

Financial investment

Access to institutions

Time and location flexibility

Without these, progress was slower.

Learning depended on:

Who you knew

Where you were

What you could afford

After A.I.

Now, much of that has shifted.

With tools like ChatGPT, individuals can:

Learn new subjects

Understand complex ideas

Build plans and strategies

Develop skills

Without waiting for access.

Knowledge Is No Longer Scarce

One of the biggest changes is this:

Knowledge is no longer limited in the same way.

It is:

Searchable

Explainable

Available

At any time.

This reduces the gap between:

Those who had access

And those who did not

Opportunity Has Expanded

A.I. allows individuals to:

Start businesses

Learn valuable skills

Create content

Solve problems

Without needing traditional pathways.

This creates more ways to move forward.

The Field Is More Open, But Not Equal

It is important to be clear:

A.I. does not eliminate all challenges.

There are still differences in:

Resources

Environment

Support systems

But what has changed is this:

The starting point is more accessible than ever before.

The New Advantage

In this new environment, advantage is no longer based only on access.

It is based on:

Awareness

Skill

Consistency

Those who:

Learn how to use A.I.

Apply it regularly

Build with it

Will move ahead.

From Access to Action

Having access is not enough.

The difference now becomes:

Who uses the tools

Who practices

Who applies what they learn

A.I. gives the opportunity.

Action creates the result.

The Role of Discipline

Even with better access, discipline still matters.

Those who:

Stay focused

Continue learning

Take consistent action

Will benefit the most.

Technology does not replace effort, it supports it.

The Risk of Inaction

While opportunity is increasing, so is the pace of change.

Those who:

Ignore A.I.

Delay learning

Avoid adapting

May find themselves behind.

Not because opportunity was unavailable, but because it was not used.

A New Way Forward

The playing field is changing.

It is becoming:

More open

More accessible

More flexible

This creates a new reality:

You are no longer limited to traditional paths.

You can:

Learn independently

Build independently

Grow independently

The Responsibility of This Moment

This moment carries responsibility.

Because with increased access comes increased expectation.

You now have:

Tools

Information

Opportunity

What you do with it matters.

Chapter Reflection

Artificial Intelligence has not made everything equal.

But it has changed what is possible.

It has created a new kind of playing field, one where:

Access is broader

Opportunity is closer

Knowledge is available

The advantage now belongs to those who:

Learn

Apply

Persist

Because in this new environment:

The difference is no longer just who had access—

it is who chooses to use it.

Chapter Fourteen

Chapter 14

Chapter 14

Hidden Doors Unlocking What Was Once Out of Reach

There are doors in this world that many people never knew existed.

Not because they were invisible,

but because they were hidden behind layers of complexity, cost, and access.

Doors to:

Knowledge

Systems

Opportunity

Understanding

For some, these doors were opened early.

For others, they remained closed, not by choice, but by limitation.

Artificial Intelligence changes that.

It does not just open doors.

It reveals doors that were once hidden.

The Nature of Hidden Doors

A hidden door is not always locked.

Sometimes it is:

Too complex to understand

Too expensive to access

Too unfamiliar to approach

These barriers have existed in areas such as:

Legal systems

Financial systems

Business strategy

Advanced education

The information was there, but not easily reachable.

A.I. as the Key

Artificial Intelligence acts as a translator, a guide, and a bridge.

It allows you to:

Take complex information and simplify it

Ask questions without restriction

Explore topics without formal entry

With tools like ChatGPT, you can now approach areas that once required specialized access.

Unlocking Knowledge in Books and Systems

Many books, documents, and systems contain valuable knowledge, but are difficult to understand.

You can now ask:

"Explain this chapter in simple terms"

"Break this down step-by-step"

"What does this mean in practical terms?"

This transforms:

Confusion → Clarity

Complexity → Understanding

Breaking Through Educational Boundaries

Education has often been limited by:

Cost

Location

Institutional access

A.I. allows you to:

Learn independently

Explore subjects deeply

Practice without restriction

This opens doors that were once tied to specific pathways.

Accessing Business and Strategy Knowledge

Business knowledge has often been reserved for:

Professionals

Consultants

Formal training environments

Now, you can ask:

"Help me build a business strategy from the ground up"

"Explain how successful businesses operate"

And begin learning immediately.

Understanding Legal and Financial Systems

Legal and financial systems are often written in language that is difficult to understand.

A.I. can help translate:

"Explain this legal document in simple terms"

"Break down this financial concept so I can understand it"

This allows individuals to:

Become more informed

Make better decisions

Navigate systems more confidently

Removing Language Barriers

Language has been another barrier.

A.I. can:

Translate languages

Simplify communication

Explain concepts across different levels of understanding

This expands access across cultures and communities.

From Information to Application

Opening a door is one thing.

Walking through it is another.

A.I. does not just provide information, it can help you apply it.

You can ask:

"How do I use this in real life?"

"What steps should I take next?"

This moves you from:

Knowing → Doing

The Power of Curiosity

Hidden doors are discovered through curiosity.

The more you ask:

The more you learn

The more you uncover

The more doors appear

A.I. responds to curiosity.

The Shift in Possibility

Before:

Access required permission

Knowledge required entry

Understanding required guidance

Now:

Access begins with a question

Knowledge is available

Guidance is immediate

This changes what is possible.

Responsibility in Accessing Hidden Doors

With greater access comes responsibility.

You must:

Use information wisely

Verify what you learn

Apply knowledge ethically

Because access without responsibility can lead to misuse.

Your Role in This New Access

You are no longer limited to what is presented to you.

You can:

Seek

Ask

Explore

Build

The doors are no longer hidden in the same way.

They are now visible, and accessible.

Chapter Reflection

Artificial Intelligence does more than provide answers.

It reveals what was once difficult to see.

It unlocks knowledge that was once out of reach.

It removes barriers that once slowed progress.

But unlocking a door is only the beginning.

The real change happens when you:

Step forward

Explore

Apply what you discover

Because in this new environment:

The doors are open.

The knowledge is available.

The next move is yours.

Chapter Fifteen

Chapter 15

Chapter 15

The Responsibility of Power Using A.I. with Discipline and Purpose

Power, by itself, is neutral.

It can build, or it can destroy.

It can uplift, or it can mislead.

It can create opportunity, or create harm.

Artificial Intelligence is power.

Not because it replaces human ability, but because it amplifies it.

Whatever you bring to A.I., it will multiply.

And that is why responsibility matters.

Understanding the Nature of Power

A.I. gives you access to:

Information

Speed

Capability

But access alone is not enough.

Without discipline, power becomes:

Misused

Wasted

Harmful

With discipline, power becomes:

Productive

Purposeful

Transformational

The Responsibility of the User

A.I. does not decide how it is used.

You do.

You choose:

What you ask

How you apply it

Whether you use it responsibly

This makes the user, not the system, the most important factor.

Using A.I. for Growth, Not Shortcuts

One of the greatest temptations is to use A.I. as a shortcut.

To:

Avoid thinking

Skip effort

Take the easy path

But shortcuts often lead to:

Weak understanding

Poor results

Limited growth

Instead, A.I. should be used to:

Strengthen thinking

Improve understanding

Increase efficiency

The Discipline to Think for Yourself

A.I. can provide answers.

But it should not replace your judgment.

You must:

Question what you receive

Verify important information

Apply your own reasoning

This keeps you in control.

Ethical Use of A.I.

With powerful tools comes ethical responsibility.

Use A.I. to:

Educate

Build

Support

Avoid using it to:

Mislead

Manipulate

Harm others

Your integrity matters more than the tool.

Avoiding Misinformation

A.I. is not perfect.

It can:

Make mistakes

Provide incomplete information

Reflect bias

This is why you must:

Cross-check important facts

Use multiple sources when necessary

Stay aware

Awareness protects you.

Balance, Not Dependence

A.I. should be part of your process, not your entire process.

Use it to:

Assist

Guide

Support

But not to:

Replace your thinking

Replace your effort

Balance creates strength.

Teaching Others Responsibly

As you learn, you may teach others.

When you do:

Encourage understanding

Promote responsible use

Share knowledge accurately

This ensures that growth spreads in a positive way.

The Long-Term Perspective

Technology will continue to evolve.

A.I. will:

Improve

Expand

Become more integrated into daily life

Your responsibility is to:

Adapt

Learn

Stay disciplined

This keeps you ahead—not overwhelmed.

Power Without Discipline Is Dangerous

Without discipline:

Knowledge becomes confusion

Access becomes distraction

Tools become misused

With discipline:

Knowledge becomes understanding

Access becomes opportunity

Tools become assets

The Choice You Have

You now have:

Access

Tools

Opportunity

What you do with them is your decision.

You can:

Use A.I. casually

Use A.I. carelessly

Or use A.I. strategically and responsibly

A.I. as a Tool for Purpose

The greatest use of A.I. is not just personal gain, it is purposeful growth.

Use it to:

Improve your life

Strengthen your community

Build something meaningful

Chapter Reflection

Artificial Intelligence is not just a tool—it is a responsibility.

It amplifies:

Your thinking

Your actions

Your intentions

Used without discipline, it can lead to confusion.

Used with discipline, it can lead to growth.

The power is real.

The access is real.

The opportunity is real.

But the outcome depends on one thing:

How you choose to use it.

Final Thought

You now understand:

What A.I. is

How to use it

How to benefit from it

And how to approach it responsibly

The knowledge is no longer hidden.

The tools are no longer out of reach.

Now the decision is yours.

Will you use this power with purpose, or let it pass by unused?

Conclusion

This is not the end.

It is the beginning.

Appendix

APPENDICES

Appendix A

The Afrodescendant A.I. Power Vault

50+ High-Impact Prompts to Turn Knowledge Into Power

This section provides practical, ready-to-use prompts designed to help you take immediate action using Artificial Intelligence. These prompts are structured to support real-life applications in business, education, personal development, and more.

Use them. Modify them. Build from them.

SECTION 1 BUSINESS & WEALTH CREATION

"Create a step-by-step business plan for a small business with a low startup cost."

"Give me 10 realistic side hustles I can start immediately."

"Create a marketing plan for my business with little to no budget."

"Help me build a brand identity for my business."

"Give me pricing strategies for my product or service."

SECTION 2 EDUCATION & LEARNING

"Explain this topic in simple terms like I'm a beginner."

"Create a 7-day study plan to learn this subject."

"Summarize this information into key points."

"Test me on this topic with questions and answers."

"Teach me this concept step-by-step."

SECTION 3 MONEY & FINANCE

"Create a simple budget based on my monthly income."

"Explain how to improve my credit step-by-step."

"Give me a plan to build wealth from zero."

"Explain investing in simple terms."

"Give me financial habits that lead to long-term success."

SECTION 4 LEGAL & STRUCTURE

"Explain this legal document in simple terms."

"How do I start an LLC step-by-step?"

"What are the basic legal protections for a business?"

"How do I prepare for a court appearance?"

"Break down my rights in simple language."

SECTION 5 HEALTH & DISCIPLINE

"Create a weekly meal plan for healthy living."

"Give me a workout plan for a busy schedule."

"Help me build a daily routine for discipline and focus."

"Give me ways to reduce stress and improve mental clarity."

"Create habits that improve energy and productivity."

SECTION 6 FAMILY & COMMUNITY

"Give me ways to improve communication in relationships."

"How can I teach discipline and confidence to my child?"

"Give me steps to resolve conflict effectively."

"How do I build a strong community group?"

"How can I mentor others effectively?"

SECTION 7 CONTENT & DIGITAL CREATION

"Give me 10 ideas for social media content."

"Write a script for a motivational video."

"Help me outline a book on [topic]."

"Create a brand message for my audience."

"Give me strategies to grow my social media."

SECTION 8 CAREER & SKILLS

"Create a professional resume based on my experience."

"Give me common interview questions and answers."

"Help me transition into a new career."

"What skills should I learn to make money today?"

"How do I start freelancing with no experience?"

SECTION 9 MINDSET & PERSONAL GROWTH

"Help me identify my strengths and purpose."

"Give me steps to build confidence."

"How do I develop a disciplined mindset?"

"Help me eliminate distractions and stay focused."

"Give me self-reflection questions to improve my life."

SECTION 10 ADVANCED STRATEGY

"Break down this situation and give me the best strategy."

"Analyze this problem and give solutions with pros and cons."

"Help me build a system that generates consistent results."

"Give me negotiation strategies for better outcomes."

"Teach me how to lead and influence effectively."

MASTER PROMPT FORMULA

Use this structure to create powerful prompts:

"Act as a [role]. Help me [goal]. Provide step-by-step instructions tailored to [my situation]."

Example:

"Act as a business consultant. Help me start a small business with a $5,000 budget. Provide step-by-step instructions with marketing strategies."

Closing Thought

What was once restricted is now accessible.

The difference is no longer access.

The difference is action.

Appendix B

Key Definitions Understanding the Language of A.I.

Artificial Intelligence (A.I.)

Technology that allows machines to perform tasks that normally require human thinking.

Prompt

The question or instruction given to an A.I. system.

Machine Learning

A process where A.I. learns from data and improves over time.

Data

Information used by A.I. to learn and generate responses.

Algorithm

A set of instructions used by a computer to solve problems.

Automation

Using technology to complete tasks automatically.

Chatbot

An A.I. system that communicates through conversation, such as ChatGPT.

Natural Language Processing (NLP)

A.I.'s ability to understand and respond to human language.

Bias

Unfair or inaccurate patterns reflected in A.I. systems based on data.

Digital Product

A product sold online, such as an e-book or course.

Freelancing

Working independently and offering services.

Content Creation

Producing digital material for an audience.

Platform

A digital space where users interact (apps, websites).

Digital Literacy

The ability to use and understand technology effectively.

Output

The result generated by A.I. based on input.

Verification

Checking information for accuracy. As if clay, shape it, mold it into your desired results.

Final Note

This appendix is not just a reference, it is a tool. Use it!!!

Return to it.

Use it.

Apply it.

Because understanding leads to confidence.

And confidence leads to action.

Afterword

Special Section

A.I. and the Future of Afrodescendant Unity, Building Across the Diaspora

Artificial Intelligence is more than a tool for individuals.

It is a tool for nations, movements, and collective advancement.

For the Afrodescendant community, spread across continents, cultures, and histories, A.I. presents a unique opportunity:

The ability to connect, organize, and build at a level that was once difficult, slow, or restricted.

This is not just about technology.

This is about unity, coordination, and forward movement.

A Global People, A Shared Opportunity

Afrodescendants exist across:

The United States

The Caribbean

South America

Africa

Europe

Connected by history, but often separated by:

Geography

Language

Systems

Access

A.I. has the ability to reduce these separations.

It allows for:

Instant communication

Language translation

Shared learning

Coordinated strategy

This creates the foundation for something powerful:

A connected global community with shared tools and shared direction

Building Knowledge Systems for the Nation

One of the greatest strengths of A.I. is its ability to organize and distribute knowledge.

The Afrodescendant Nation can use A.I. to:

Develop educational materials

Create training programs

Preserve history and cultural knowledge

Build structured learning systems

Instead of knowledge being scattered, it can be:

Centralized

Simplified

Shared across all members

Economic Empowerment Through A.I.

A unified community must also be economically strong.

A.I. can support:

Business development

Market analysis

Cooperative economics

Digital marketplaces

Entrepreneurs across the diaspora can:

Share strategies

Build partnerships

Support each other's businesses

This strengthens economic independence.

Communication and Coordination

A.I. allows for improved communication within and across communities.

It can help:

Draft messages and announcements

Translate languages instantly

Organize information clearly

This allows organizations to:

Operate more efficiently

Reach more people

Maintain clarity in communication

Education Without Borders

A.I. removes the need for physical location in learning.

Members across the diaspora can:

Access the same information

Learn the same systems

Develop the same skills

This creates:

Standardized knowledge across a global community

Training the Next Generation

Youth are the future of any nation.

A.I. allows for:

Early exposure to technology

Skill development

Leadership training

Young people can be taught:

How to use A.I. responsibly

How to build with it

How to lead with knowledge

This ensures continuity and growth.

Preserving and Sharing Cultural Identity

A.I. can also be used to:

Document history

Preserve traditions

Share cultural narratives

This ensures that identity is:

Maintained

Understood

Passed forward

Strategic Planning and Nation-Building

A.I. can assist leadership in:

Planning initiatives

Analyzing challenges

Developing strategies

It provides:

Structure

Insight

Organized thinking

While it does not replace leadership, it strengthens it.

Unity Through Shared Tools

Unity is not only about belief, it is about function.

When a community uses:

The same tools

The same systems

The same methods

It becomes more aligned.

A.I. provides a common platform for:

Learning

Building

Communicating

Responsibility in Collective Use

With collective power comes collective responsibility.

The use of A.I. within the community must be:

Ethical

Accurate

Purpose-driven

It should be used to:

Build, not divide

Educate, not mislead

Strengthen, not weaken

A New Possibility for Unity

In the past, unity required:

Physical gatherings

Slow communication

Limited coordination

Today, unity can be:

Immediate

Organized

Scalable

A.I. makes it possible to:

Connect across borders

Share across systems

Build across generations

Your Role in This Movement

You do not need to wait for a large system to begin.

You can:

Learn A.I.

Share it with others

Apply it within your community

Contribute to collective growth

Small actions lead to larger change.

Final Reflection

Artificial Intelligence is not just advancing technology.

It is expanding possibility.

For the Afrodescendant Nation and the global Afrodescendant community, it offers:

Connection

Coordination

Growth

Empowerment

But like all tools, its impact depends on how it is used.

Used individually, it can improve lives.

Used collectively, it can strengthen a people.

The opportunity is present.

The tools are available.

The question is:

Will we use them together?

About the author

Naaji Mansoor is a lifelong member of the Afrodescendant community and a dedicated advocate for self-determination, economic empowerment, and collective advancement. Based in Boston, Massachusetts, his work reflects a deep commitment to empowering individuals and strengthening communities through knowledge, structure, and purposeful action.

From a young age, he joined the Lost-Found Nation of Islam under the direction of The Honorable Silis Muhammad, where he began developing his foundation in discipline, purpose, and service.

Naaji Mansoor currently serves as a member of the House of Representatives within the Afrodescendant Nation (ADN), contributing to initiatives focused on governance, community development, and strategic nation-building.

As an author and businessman, he is committed to creating practical tools, knowledge systems, and solutions that uplift individuals and communities alike. His work centers on advancing reparatory justice, strengthening economic independence, and promoting unity through education and structured empowerment.

Driven by a clear mission, Naaji Mansoor continues to pursue and advocate for freedom, justice, and equality for all, using his voice, platforms, and publications to inspire action and lasting change.

Contact

For inquiries, speaking engagements, collaborations, or business opportunities:

Email: NaajiMM@yahoo.com